The BURNER Journal

Your burning desire is the key to achieving.

Created by Tristan Reynolds

Ashland Mills
PUBLISHING

Published by Ashland Mills Publishing
ISBN 978-0-9998722-0-8

E.C.C.

PRINTED IN AMERICA
Environmentally Conscious Company

Start by signing the cover of your journal in the white area below The BURNER Journal.

My legally binding commitment

I ______________________ have committed to following through to the end of this journal. I am committed to improving my life through practice.

What's inside?

Three self exploration
Learn about yourself

Three months of mindfulness
Practice mindfulness daily

Daily fortune cookie
Treat your mind and soul

Morning and evening check-in
Check how your feeling daily

Twelve thoughts
Once a week question or small adventure plan

In ***The Burner Journal*** we explore your deepest ***burning desires*** to find new practice while practicing ***mindfulness*** and creating stronger abilities to be more ***self-aware***. Mindfulness is ***a practice of overcoming fears, let go of inadequacies, and excepting the things you can not change.*** When you are mindful with focused energy you add a skip in your step by knowing when it's safe to have playful adventures and when it is time to focus. Transforming positive habits into a lifestyle not just a temparay diet.

Gently take small steps leaving footprints in the sand to be the best version of yourself through this practice and everyone around you will become better in your presence. *"What we call "i" is just a swinging door which moves when we inhale and when we exhale" Stop, breath and form little attachment to things.*

"We should find perfect existence through imperfect exsistence."

"If you lose the spirit of repetion, your practice will become quite difficult." Shunryu Suzuki

Let go of perfection and simplely aim to improve. Refocus your energy on a perfectly lifestyle for as as imperfect human. Taking the best care of yourself by simply doing everyday things such as eating in the morning and night creates the ablity to help others as an extra. By not worring about taking credit of your all your extra you will find a beuatiful exsistance.

On this adventure of being the best version of yourself you will find you have to practice self care, mindfullness and small adventures to achieve a happier life.

SAY
WHAT
EVER
YOU
NEED
TO
HOW
EVER
YOU
NEED
TO

Write as though you are whispering to someone important.

I want to be better at...

"Always concentrate on how far you've come, rather than how far you have left to go."

Write down what you wish to have in your life in the next couple years...

Now take a short walk before moving on to the next page.

TALK
ABOUT
WHAT
EVER
YOU
WANT
TO
HOW
EVER
YOU
WANT
TO

Write as though you're standing on a mountain top shouting.

My burning desires are...

Smile
MORE
worry
LESS

Write as though you are in the moment.

Three things that make me smile are...

The Paper Airplane Challenge

2nd Fold
(Fold Line)

3rd Fold
(Fold Line)

Write what I think the answer to a most recent fear(s), or obstacle(s) is...

Tear this page out then fold it into a paper airplane.

Tear Page out ✂

4th Fold Line

1st Fold Line

5th Fold Line

Throw the plane releasing it away from you and others.

Pick up the plane, crumple it into a ball and throw it away.

What should I be grateful for? Everything around you has some form of besuty the moment with a friend, a meal, a tree, rain, snow, sun and even the smallest bug.

One can be more energized by simply starting or stopping. Sometimes simply stopping to breath and starting a walk can bring energy.

Affirmations, You are great everyday even when you feel you are not. You may have more strength then you are aware.

The daily fortune cookie is a treat for your mind. It is a fortune, humorous or a important quote.

Morning and Evening feelings checkin
By checking in with yourself it makes you more self-ware of what you are gratful for, what energizes you or brings a small victory that energized you enough to start a new practice of be your best self.

It is important to know how far you have come and note the small victories.

How a challenge is important to you has great value in finding your best direction forward.
What you need or someone important to you needs can shows self care and empathy.

TIME 7:15 — **M** T W T F S S — DATE: 8/5/18

TODAY, I AM GRATEFUL FOR... The sunny day today and the conversation I had with my friend Tom.

TO FEEL MORE ENERGIZED I WILL... Walk everyday to clear my mind

AFFIRMATIONS, I AM... a good brother to my siblings.
I am a great friend.
I am a kind person.

Nothing has to be perfect.

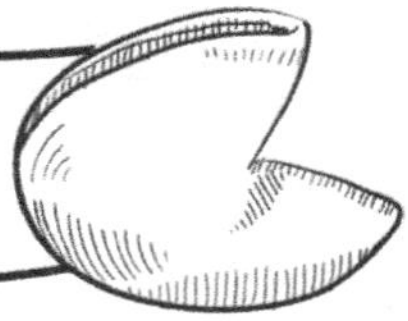

This morning I feel.
(Circle one below)

Displeased — Fair — **Exceptional**

Displeased — **Fair** — Exceptional

Tonight I feel.
(Circle one above)

One small victory for today was... I walked a mile today with my friend Mike and we planned a daily walk.

What did I takeaway from today's challenge(s)? I learned that I need to be more focused when people talk and try to do my best to understand a situation before I judge.

TIME M T W T F S S DATE:

TODAY, I AM GRATEFUL FOR... ______

______

______

TO FEEL MORE ENERGIZED I WILL... ______

AFFIRMATIONS, I AM... ______

______

______

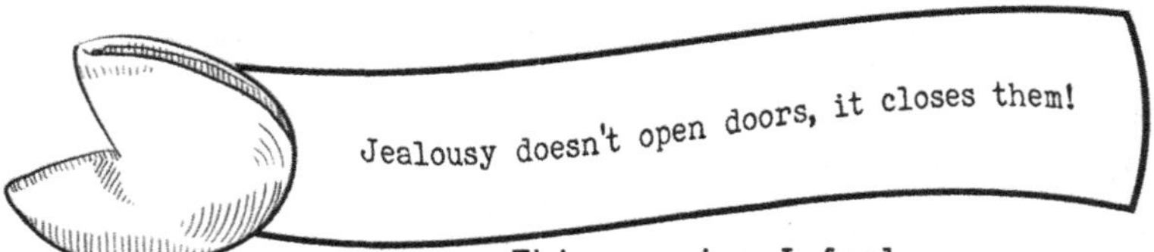

This morning I feel.
(Circle one below)

Fair

Displeased Exceptional

Displeased Exceptional

Fair

Tonight I feel.
(Circle one above)

One small victory for today was... ______

______

What did I takeaway from today's challenge(s)? ______

______

______

TIME M T W T F S S DATE:

TODAY, I AM GRATEFUL FOR... ______________________________

____

TO FEEL MORE ENERGIZED I WILL... ______________________________

AFFIRMATIONS, I AM... ______________________________

____

____

Help, I'm prisoner in a fortune cookie factory!!!

This morning I feel.
(Circle one below)

Fair

Displeased

Exceptional

Displeased

Exceptional

Fair

Tonight I feel.
(Circle one above)

One small victory for today was... ______________________________

____

What did I takeaway from today's challenge(s)? ______________________________

____

____

TIME M T W T F S S DATE:

TODAY, I AM GRATEFUL FOR... ______________________

____

____

TO FEEL MORE ENERGIZED I WILL... ______________________

AFFIRMATIONS, I AM... ______________________

____

____

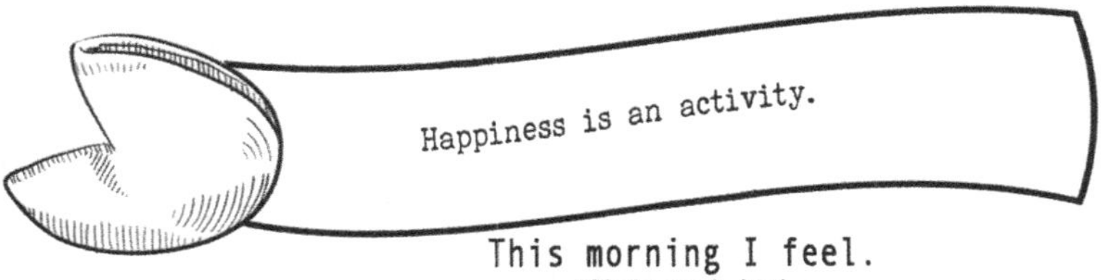

This morning I feel.
(Circle one below)

Fair

Displeased

Exceptional

Displeased

Exceptional

Fair

Tonight I feel.
(Circle one above)

One small victory for today was... ______________________

____

What did I takeaway from today's challenge(s)? ______________________

____

____

TIME M T W T F S S DATE:

TODAY, I AM GRATEFUL FOR... ______

______

TO FEEL MORE ENERGIZED I WILL... ______

AFFIRMATIONS, I AM... ______

______

______

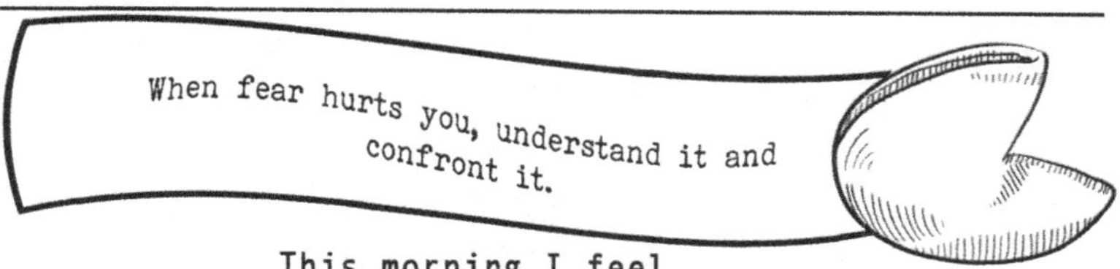

This morning I feel.
(Circle one below)

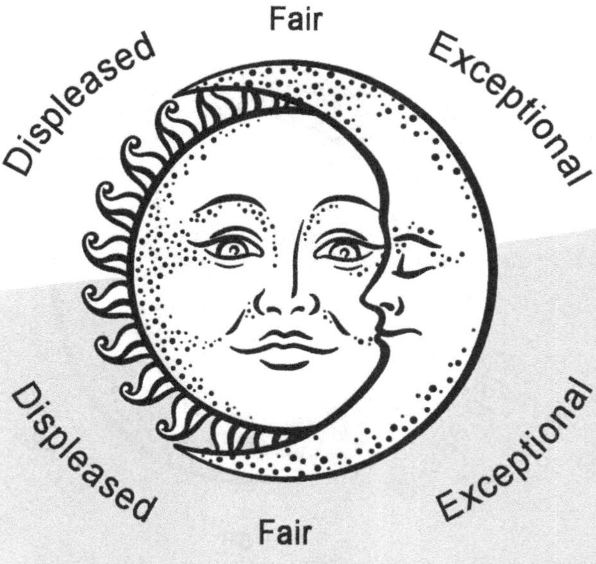

Tonight I feel.
(Circle one above)

One small victory for today was... ______

______

How can I make tomorrow better for myself and/or others? ______

______

______

TIME M T W T F S S DATE:

TODAY, I AM GRATEFUL FOR... ______

TO FEEL MORE ENERGIZED I WILL... ______

AFFIRMATIONS, I AM... ______

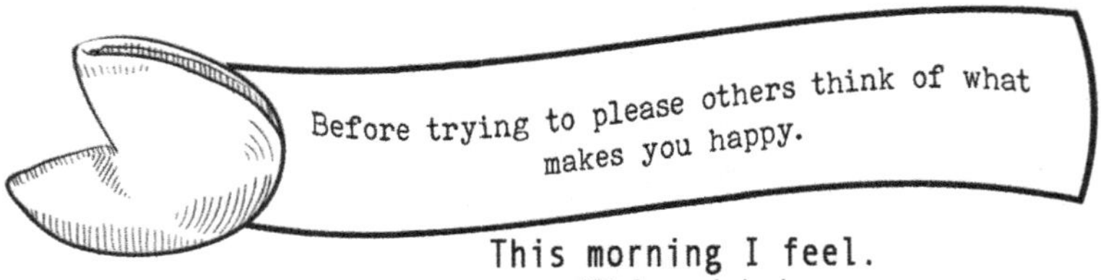

This morning I feel.
(Circle one below)

Fair

Displeased

Exceptional

Displeased

Exceptional

Fair

Tonight I feel.
(Circle one above)

One small victory for today was... ______

What is one small thing I can do for someone else next week?

You can go anywhere

Shade in a place you would like to visit.

TIME M T W T F S S DATE:

TODAY, I AM GRATEFUL FOR... ______________________________
____
____

TO FEEL MORE ENERGIZED I WILL... ______________________________

AFFIRMATIONS, I AM... ______________________________
____
____

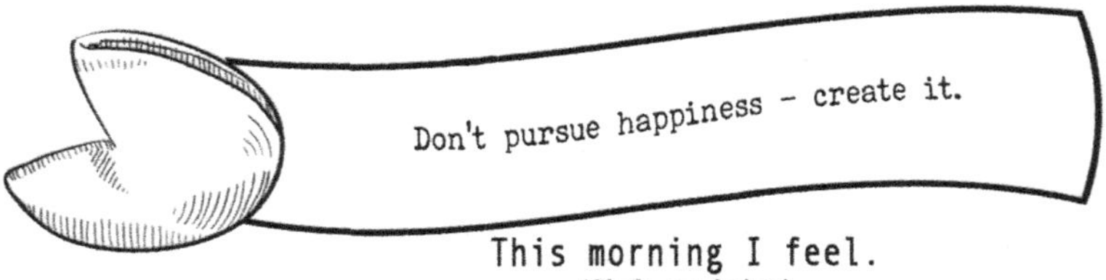

This morning I feel.
(Circle one below)

Fair

Displeased

Exceptional

Displeased

Exceptional

Fair

Tonight I feel.
(Circle one above)

One small victory for today was... ______________________________
____

What did I takeaway from today's challenge(s)? ______________________________
____
____

TIME M T W T F S S DATE:

TODAY, I AM GRATEFUL FOR...

TO FEEL MORE ENERGIZED I WILL...

AFFIRMATIONS, I AM...

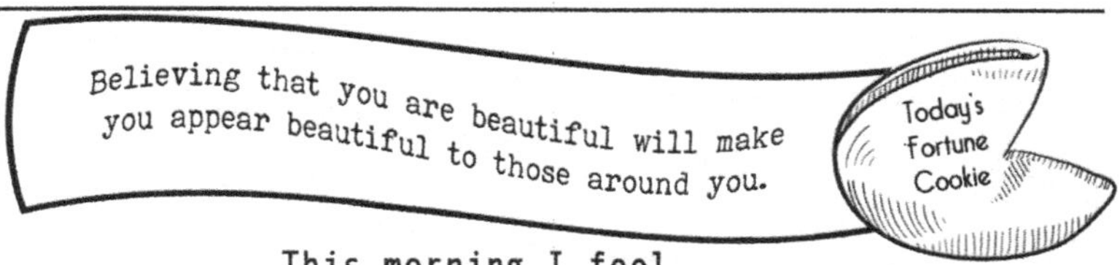

This morning I feel.
(Circle one below)

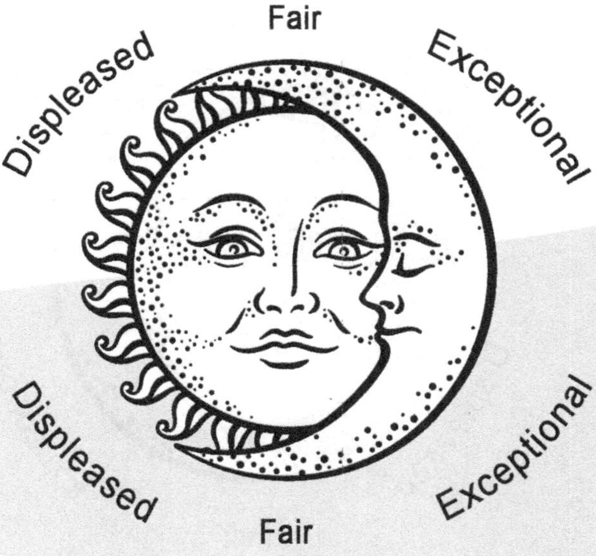

Tonight I feel.
(Circle one above)

One small victory for today was...

What did I takeaway from today's challenge(s)?

TIME M T W T F S S DATE:

TODAY, I AM GRATEFUL FOR... ______

TO FEEL MORE ENERGIZED I WILL... ______

AFFIRMATIONS, I AM... ______

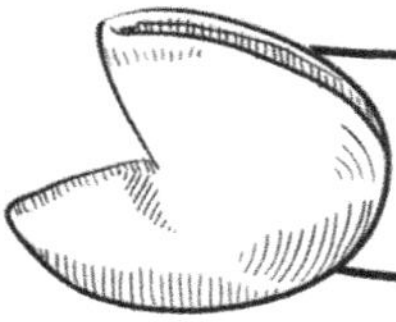

The time is always right to do what is right.

This morning I feel.
(Circle one below)

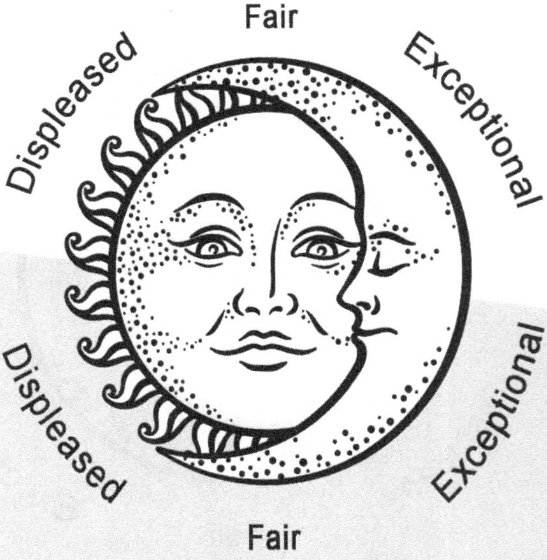

Tonight I feel.
(Circle one above)

One small victory for today was... ______

What did I takeaway from today's challenge(s)? ______

TIME M T W T F S S DATE:

TODAY, I AM GRATEFUL FOR... ______________________________

____

TO FEEL MORE ENERGIZED I WILL... ______________________________

AFFIRMATIONS, I AM... ______________________________

____

____

If your desires are not extravagant, they will be rewarded

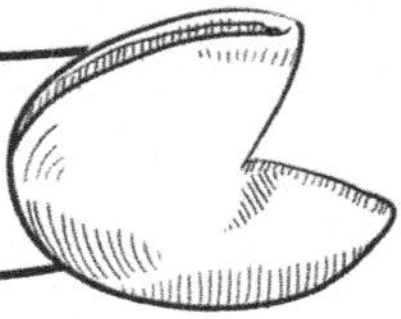

This morning I feel.
(Circle one below)

Displeased Fair Exceptional

Displeased Fair Exceptional

Tonight I feel.
(Circle one above)

One small victory for today was... ______________________________

____

How can I make tomorrow better for myself and/or others? ______________________________

____

____

TIME M T W T F S S DATE:

TODAY, I AM GRATEFUL FOR...

TO FEEL MORE ENERGIZED I WILL...

AFFIRMATIONS, I AM...

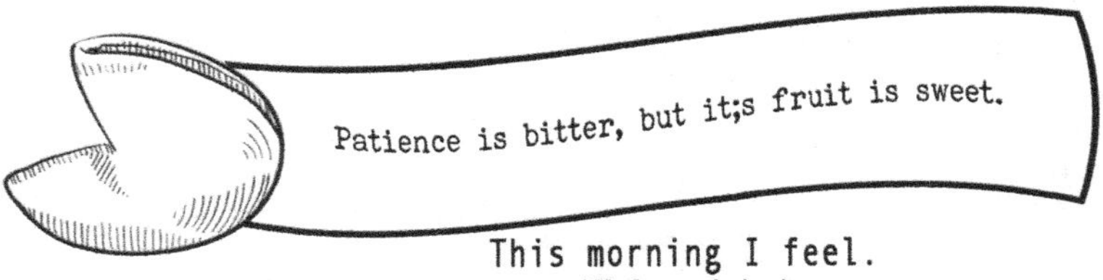

This morning I feel.
(Circle one below)

Fair

Displeased Exceptional

Displeased Exceptional

Fair

Tonight I feel.
(Circle one above)

One small victory for today was...

What is one small thing I can do for someone else next week?

You can go anywhere

Shade in a place you would like to visit.

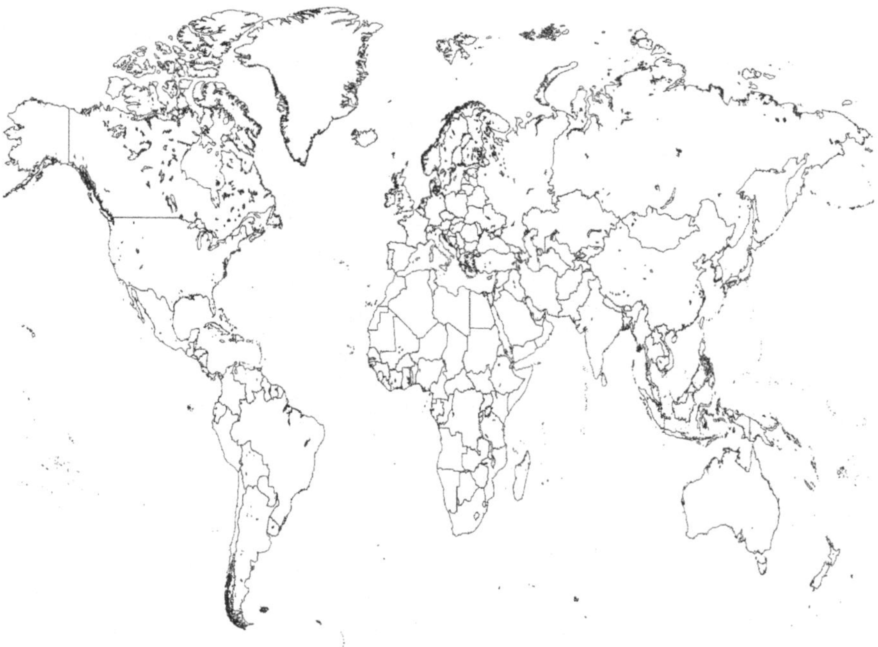

TIME M T W T F S S DATE:

TODAY, I AM GRATEFUL FOR... ______________________________

____

____

TO FEEL MORE ENERGIZED I WILL... ______________________________

AFFIRMATIONS, I AM... ______________________________

____

____

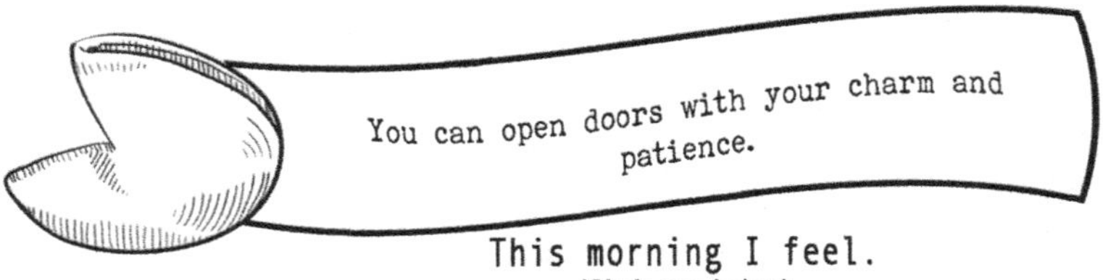

This morning I feel.
(Circle one below)

Fair

Displeased

Exceptional

Displeased

Exceptional

Fair

Tonight I feel.
(Circle one above)

One small victory for today was... ______________________________

____

What did I takeaway from today's challenge(s)? ______________________________

____

____

TIME M T W T F S S DATE:

TODAY, I AM GRATEFUL FOR...

TO FEEL MORE ENERGIZED I WILL...

AFFIRMATIONS, I AM...

Nothing has to be perfect.

Today's Fortune Cookie

This morning I feel.
(Circle one below)

Displeased Fair Exceptional

Displeased Fair Exceptional

Tonight I feel.
(Circle one above)

One small victory for today was...

What did I takeaway from today's challenge(s)?

TIME M T W T F S S DATE:

TODAY, I AM GRATEFUL FOR... ______

TO FEEL MORE ENERGIZED I WILL... ______

AFFIRMATIONS, I AM... ______

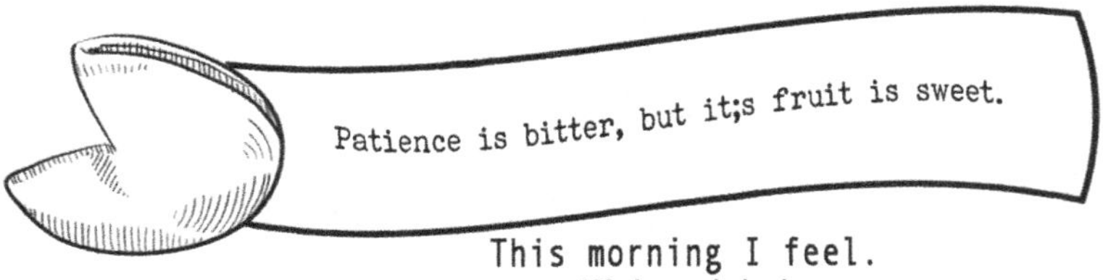

This morning I feel.
(Circle one below)

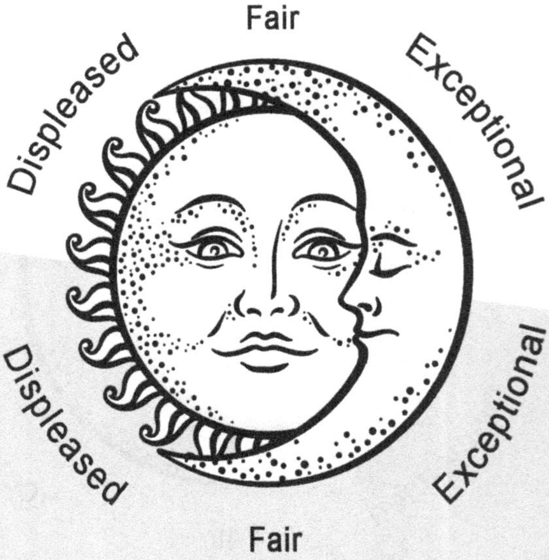

Tonight I feel.
(Circle one above)

One small victory for today was... ______

What did I takeaway from today's challenge(s)? ______

TIME M T W T F S S DATE:

TODAY, I AM GRATEFUL FOR... ____________________

____

TO FEEL MORE ENERGIZED I WILL... ____________________

AFFIRMATIONS, I AM... ____________________

____

____

Believing that you are beautiful will make you appear beautiful to those around you.

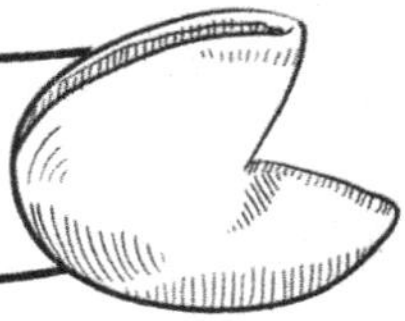

This morning I feel.
(Circle one below)

Fair

Displeased

Exceptional

Displeased

Exceptional

Fair

Tonight I feel.
(Circle one above)

One small victory for today was... ____________________

____

How can I make tomorrow better for myself and/or others? ____________________

____

____

TIME M T W T F S S DATE:

TODAY, I AM GRATEFUL FOR... ______

______

______

TO FEEL MORE ENERGIZED I WILL... ______

AFFIRMATIONS, I AM... ______

______

______

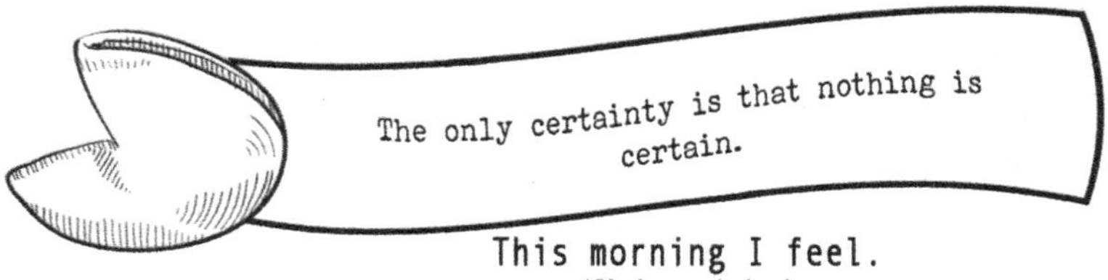

This morning I feel.
(Circle one below)

Displeased Fair Exceptional

Displeased Fair Exceptional

Tonight I feel.
(Circle one above)

One small victory for today was... ______

______

What is one small thing I can do for someone else next week?

______

Create Your Own Magic

What is one new activity I want to start this year?

TIME M T W T F S S DATE:

TODAY, I AM GRATEFUL FOR... ____________________

____

____

TO FEEL MORE ENERGIZED I WILL... ____________________

AFFIRMATIONS, I AM... ____________________

____

____

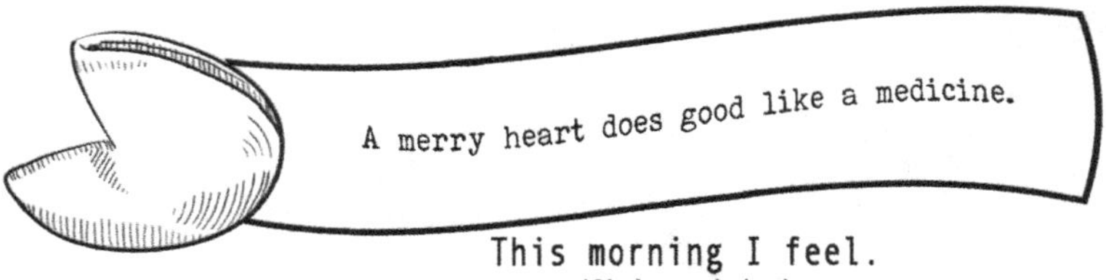

This morning I feel.
(Circle one below)

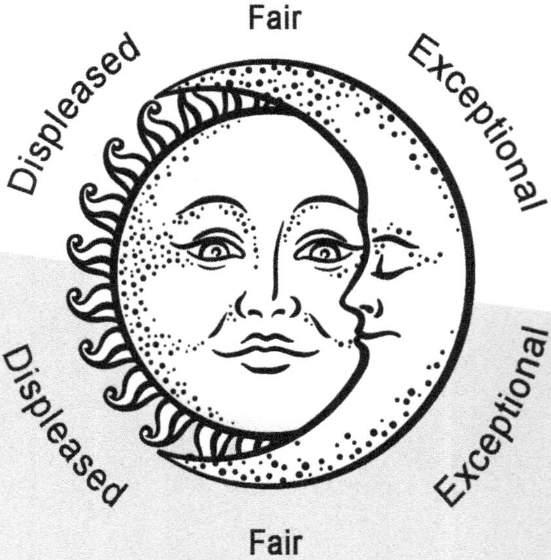

Tonight I feel.
(Circle one above)

One small victory for today was... ____________________

____

What did I takeaway from today's challenge(s)? ____________________

____

____

TIME M T W T F S S DATE:

TODAY, I AM GRATEFUL FOR... ______________________

____

TO FEEL MORE ENERGIZED I WILL... ______________________

AFFIRMATIONS, I AM... ______________________

____

____

Now these three remaim, faith, hope, and love. The greatest of these is love.

This morning I feel.
(Circle one below)

Fair

Displeased

Exceptional

Displeased

Exceptional

Fair

Tonight I feel.
(Circle one above)

One small victory for today was... ______________________

____

What did I takeaway from today's challenge(s)? ______________________

____

____

TIME M T W T F S S DATE:

TODAY, I AM GRATEFUL FOR... ______

______

______

TO FEEL MORE ENERGIZED I WILL... ______

AFFIRMATIONS, I AM... ______

______

______

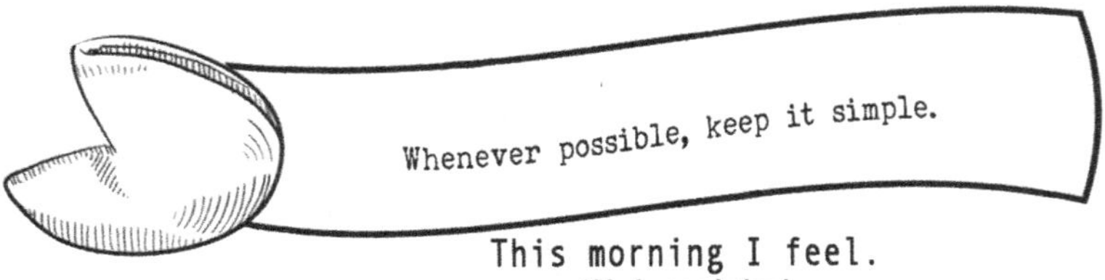

This morning I feel.
(Circle one below)

Displeased Fair Exceptional

Displeased Fair Exceptional

Tonight I feel.
(Circle one above)

One small victory for today was... ______

______

What did I takeaway from today's challenge(s)? ______

______

______

TIME M T W T F S S DATE:

TODAY, I AM GRATEFUL FOR... ______________________________

____

TO FEEL MORE ENERGIZED I WILL... ______________________________

AFFIRMATIONS, I AM... ______________________________

____

____

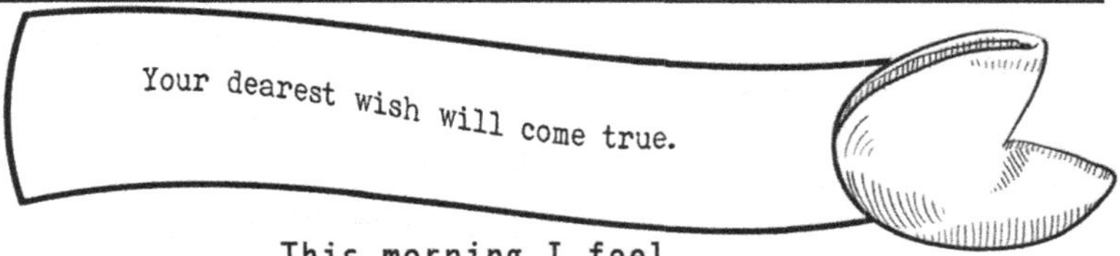

This morning I feel.
(Circle one below)

Fair

Displeased Exceptional

Displeased Exceptional

Fair

Tonight I feel.
(Circle one above)

One small victory for today was... ______________________________

____

How can I make tomorrow better for myself and/or others? ______________________________

____

____

TIME M T W T F S S DATE:

TODAY, I AM GRATEFUL FOR... ______

TO FEEL MORE ENERGIZED I WILL... ______

AFFIRMATIONS, I AM... ______

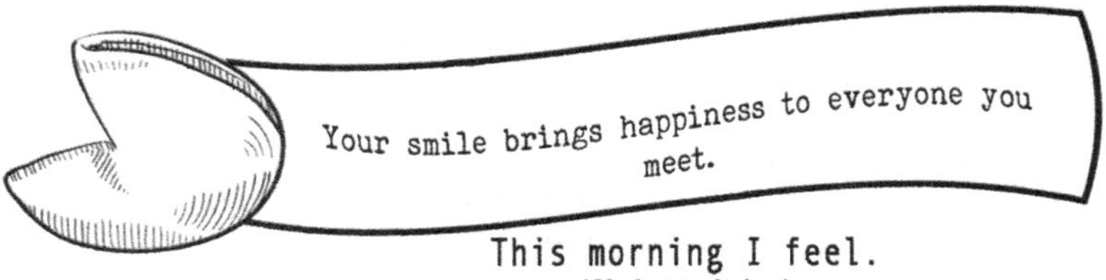

This morning I feel.
(Circle one below)

Fair

Displeased

Exceptional

Displeased

Exceptional

Fair

Tonight I feel.
(Circle one above)

One small victory for today was... ______

What is one small thing I can do for someone else next week?

without ACTION you aren't going ANYWHERE

What action are you proud of this week?

TIME M T W T F S S DATE:

TODAY, I AM GRATEFUL FOR... ______________________________

____

____

TO FEEL MORE ENERGIZED I WILL... ______________________________

AFFIRMATIONS, I AM... ______________________________

____

____

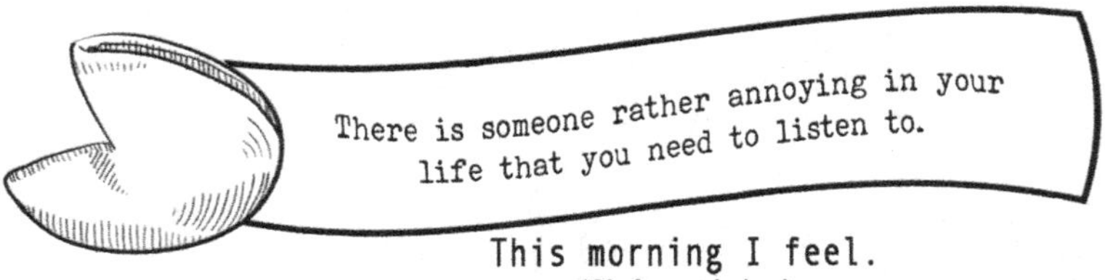

This morning I feel.
(Circle one below)

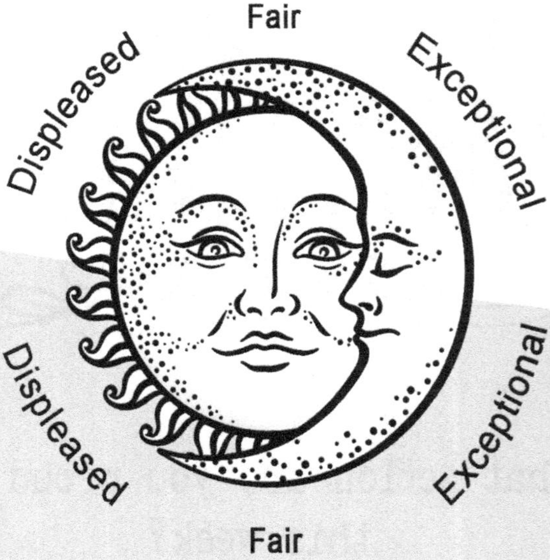

Tonight I feel.
(Circle one above)

One small victory for today was... ______________________________

____

What did I takeaway from today's challenge(s)? ______________________________

____

____

TIME M T W T F S S DATE:

TODAY, I AM GRATEFUL FOR... ____________________

____

TO FEEL MORE ENERGIZED I WILL... ____________________

AFFIRMATIONS, I AM... ____________________

____

____

Your life will prosper only if you acknowledge your faults and work to reduce them.

This morning I feel.
(Circle one below)

Fair

Displeased

Exceptional

Displeased

Exceptional

Fair

Tonight I feel.
(Circle one above)

One small victory for today was... ____________________

____

What did I takeaway from today's challenge(s)? ____________________

____

____

TIME M T W T F S S DATE:

TODAY, I AM GRATEFUL FOR...

TO FEEL MORE ENERGIZED I WILL...

AFFIRMATIONS, I AM...

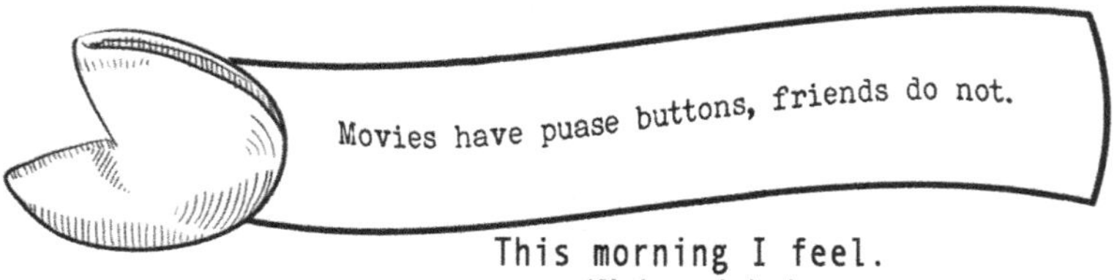

This morning I feel.
(Circle one below)

Fair

Displeased

Exceptional

Displeased

Exceptional

Fair

Tonight I feel.
(Circle one above)

One small victory for today was...

What did I takeaway from today's challenge(s)?

TIME M T W T F S S DATE:

TODAY, I AM GRATEFUL FOR... ___

___

TO FEEL MORE ENERGIZED I WILL... ___

AFFIRMATIONS, I AM... ___

___

___

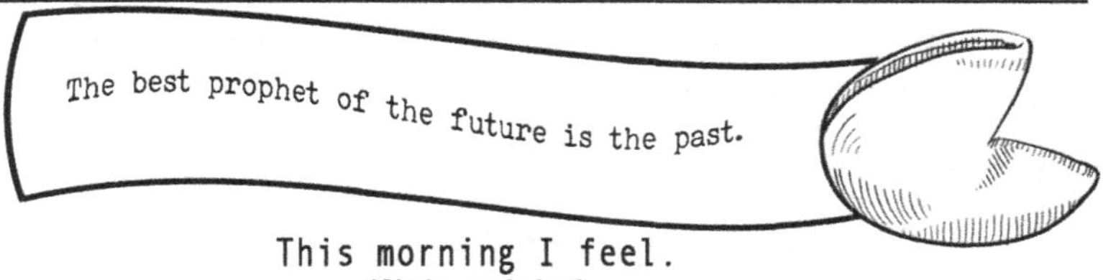

This morning I feel.
(Circle one below)

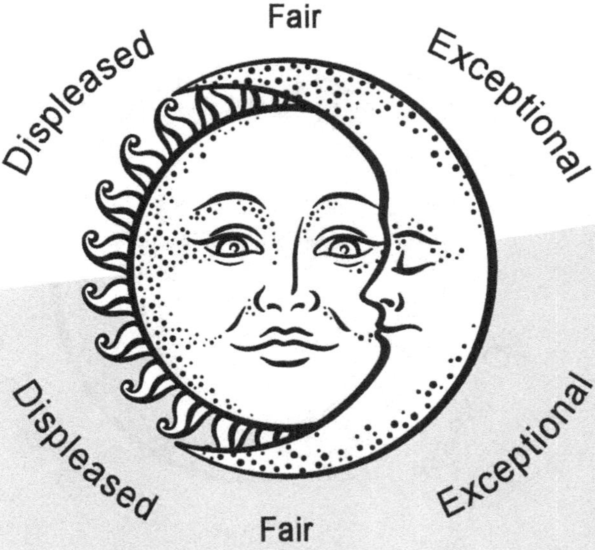

Tonight I feel.
(Circle one above)

One small victory for today was... ___

___

How can I make tomorrow better for myself and/or others? ___

___

___

TIME M T W T F S S DATE:

TODAY, I AM GRATEFUL FOR... ______________________________

____

____

TO FEEL MORE ENERGIZED I WILL... ______________________________

AFFIRMATIONS, I AM... ______________________________

____

____

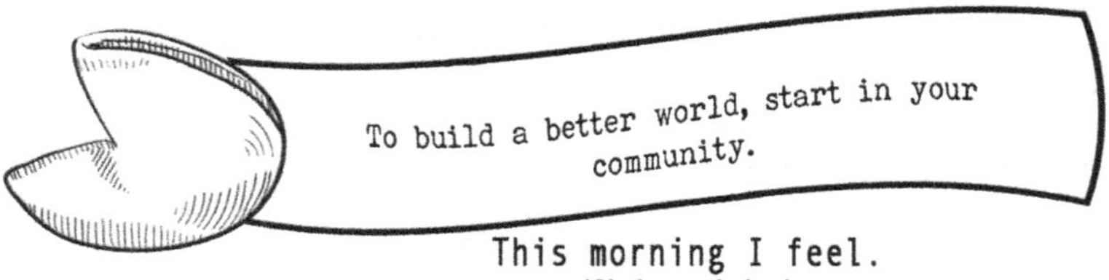

This morning I feel.
(Circle one below)

Fair

Displeased Exceptional

Displeased Exceptional

Fair

Tonight I feel.
(Circle one above)

One small victory for today was... ______________________________

____

What is one small thing I can do for someone else next week?

____

without ACTION you aren't going ANYWHERE

What action are you proud of
this week?

TIME M T W T F S S DATE:

TODAY, I AM GRATEFUL FOR... ______________________________

____

____

TO FEEL MORE ENERGIZED I WILL... ______________________________

AFFIRMATIONS, I AM... ______________________________

____

____

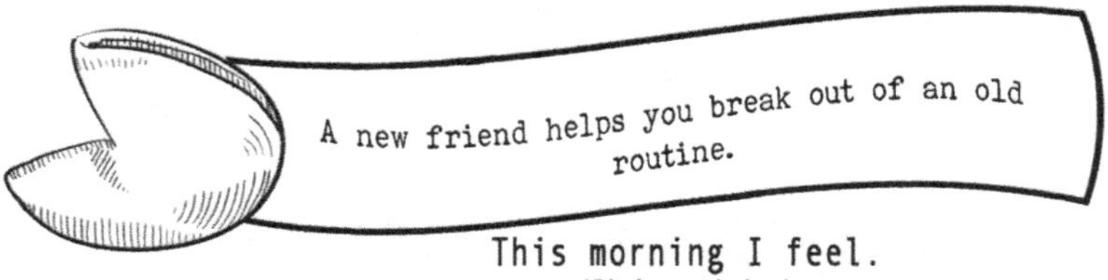

This morning I feel.
(Circle one below)

Fair

Displeased

Exceptional

Displeased

Exceptional

Fair

Tonight I feel.
(Circle one above)

One small victory for today was... ______________________________

____

What did I takeaway from today's challenge(s)? ______________________________

____

____

TIME M T W T F S S DATE:

TODAY, I AM GRATEFUL FOR... ____________________

____

TO FEEL MORE ENERGIZED I WILL... ____________________

AFFIRMATIONS, I AM... ____________________

____

____

The best is yet to come.

This morning I feel.
(Circle one below)

Displeased Fair Exceptional

Displeased Fair Exceptional

Tonight I feel.
(Circle one above)

One small victory for today was... ____________________

____

What did I takeaway from today's challenge(s)? ____________________

____

____

TIME M T W T F S S DATE:

TODAY, I AM GRATEFUL FOR... ______

______

______

TO FEEL MORE ENERGIZED I WILL... ______

AFFIRMATIONS, I AM... ______

______

______

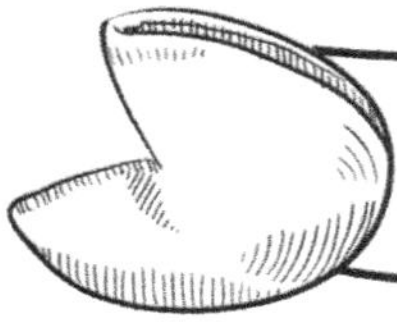

Do not be overly judgemental of your loved one's intentions or actions.

This morning I feel.
(Circle one below)

Fair

Displeased

Exceptional

Displeased

Exceptional

Fair

Tonight I feel.
(Circle one above)

One small victory for today was... ______

______

What did I takeaway from today's challenge(s)? ______

______

______

TIME M T W T F S S DATE:

TODAY, I AM GRATEFUL FOR... ______________________________

____

TO FEEL MORE ENERGIZED I WILL... ______________________________

AFFIRMATIONS, I AM... ______________________________

____

____

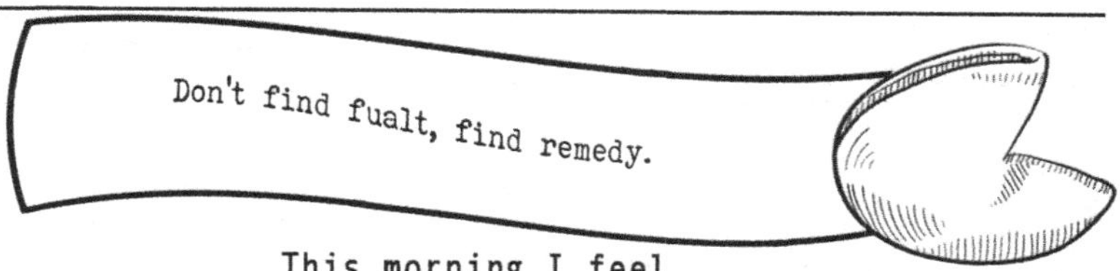

This morning I feel.
(Circle one below)

Displeased Fair Exceptional

Displeased Fair Exceptional

Tonight I feel.
(Circle one above)

One small victory for today was... ______________________________

____

How can I make tomorrow better for myself and/or others? ______________________________

____

____

TIME M T W T F S S DATE:

TODAY, I AM GRATEFUL FOR... ______________________________

____

____

TO FEEL MORE ENERGIZED I WILL... ______________________________

AFFIRMATIONS, I AM... ______________________________

____

____

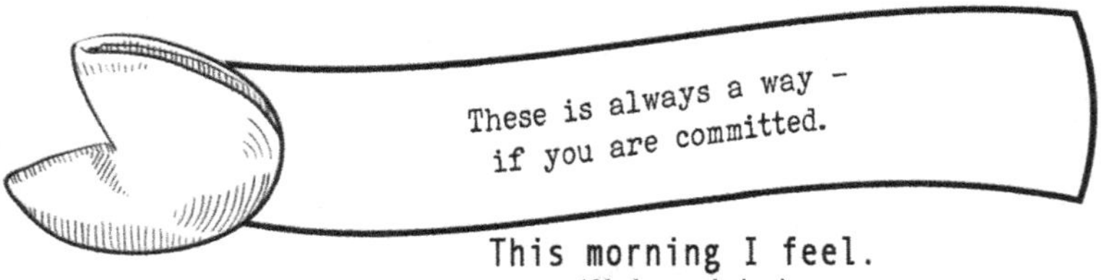

This morning I feel.
(Circle one below)

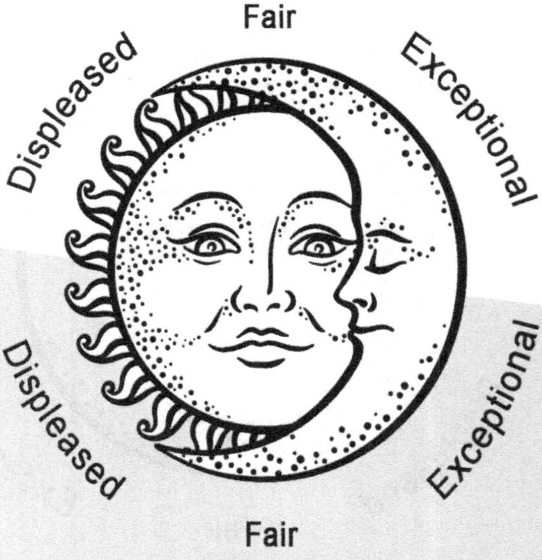

Tonight I feel.
(Circle one above)

One small victory for today was... ______________________________

____

What is one small thing I can do for someone else next week?

____

Smile
MORE
worry
LESS

TIME M T W T F S S DATE:

TODAY, I AM GRATEFUL FOR... ____________________

____

____

TO FEEL MORE ENERGIZED I WILL... ____________________

AFFIRMATIONS, I AM... ____________________

____

____

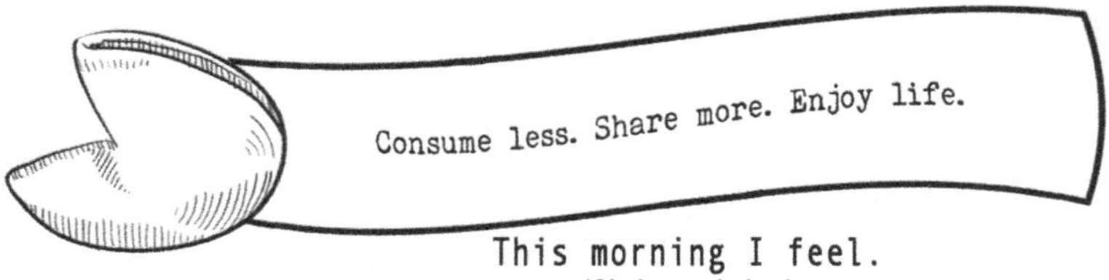

This morning I feel.
(Circle one below)

Fair

Displeased Exceptional

Displeased Exceptional

Fair

Tonight I feel.
(Circle one above)

One small victory for today was... ____________________

____

What did I takeaway from today's challenge(s)? ____________________

____

____

TIME M T W T F S S DATE:

TODAY, I AM GRATEFUL FOR... ______________________________

____

TO FEEL MORE ENERGIZED I WILL... ______________________________

AFFIRMATIONS, I AM... ______________________________

____

____

If you want the rainbow, then you have to tolerate the rain.

This morning I feel.
(Circle one below)

Displeased Fair Exceptional

Displeased Fair Exceptional

Tonight I feel.
(Circle one above)

One small victory for today was... ______________________________

____

What did I takeaway from today's challenge(s)? ______________________________

____

____

TIME M T W T F S S DATE:

TODAY, I AM GRATEFUL FOR... ______________________________

____

____

TO FEEL MORE ENERGIZED I WILL... ______________________________

AFFIRMATIONS, I AM... ______________________________

____

____

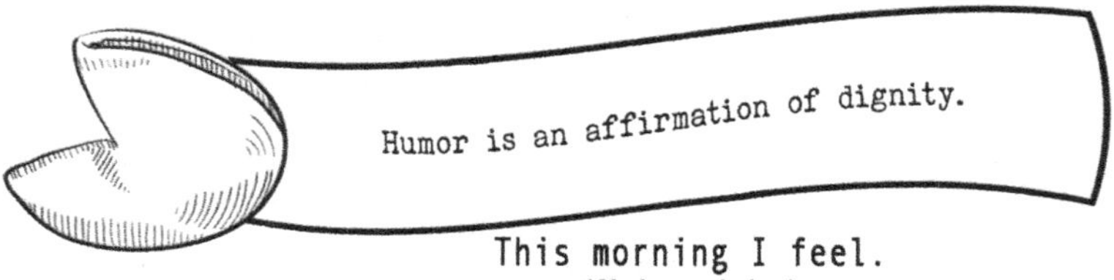

This morning I feel.
(Circle one below)

Fair

Displeased

Exceptional

Displeased

Exceptional

Fair

Tonight I feel.
(Circle one above)

One small victory for today was... ______________________________

____

What did I takeaway from today's challenge(s)? ______________________________

____

____

TIME M T W T F S S DATE:

TODAY, I AM GRATEFUL FOR... ____________________

____

TO FEEL MORE ENERGIZED I WILL... ____________________

AFFIRMATIONS, I AM... ____________________

____

____

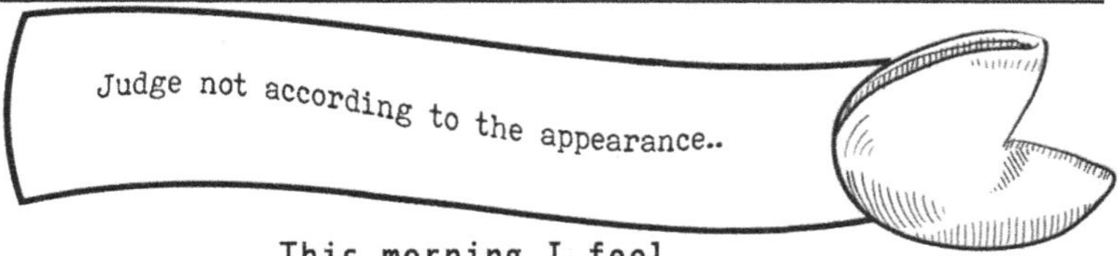

This morning I feel.
(Circle one below)

Displeased Fair Exceptional

Displeased Fair Exceptional

Tonight I feel.
(Circle one above)

One small victory for today was... ____________________

____

How can I make tomorrow better for myself and/or others? ____________________

____

____

TIME M T W T F S S DATE:

TODAY, I AM GRATEFUL FOR...

TO FEEL MORE ENERGIZED I WILL...

AFFIRMATIONS, I AM...

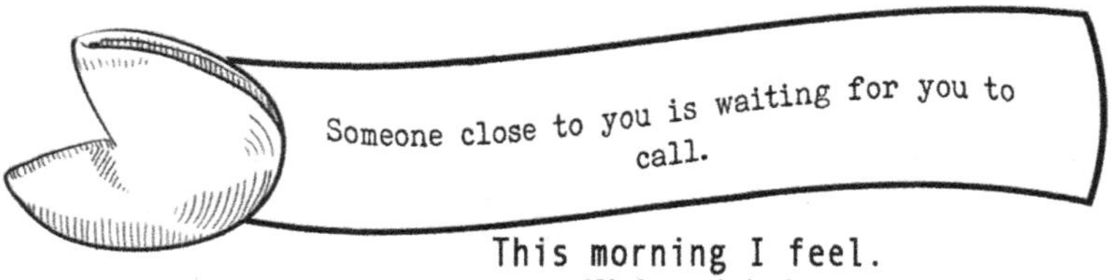

This morning I feel.
(Circle one below)

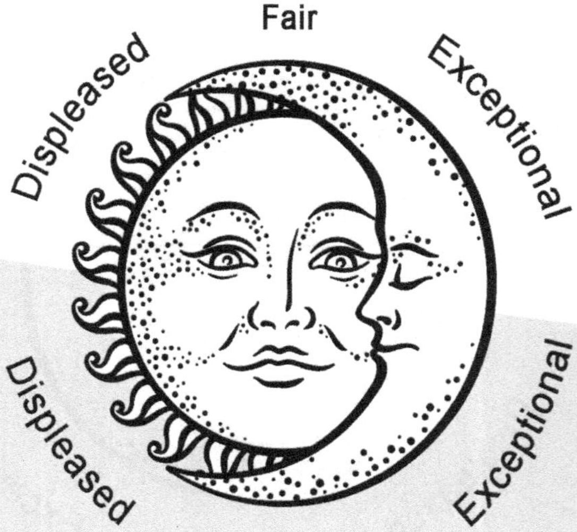

Tonight I feel.
(Circle one above)

One small victory for today was...

What is one small thing I can do for someone else next week?

the adventure begins

Write a small adventure you would like to take using the questions below.

Where?

When?

How to make it happen?

What do I need?

Who will go with?

TIME M T W T F S S DATE:

TODAY, I AM GRATEFUL FOR... ______________________________

____

____

TO FEEL MORE ENERGIZED I WILL... ______________________________

AFFIRMATIONS, I AM... ______________________________

____

____

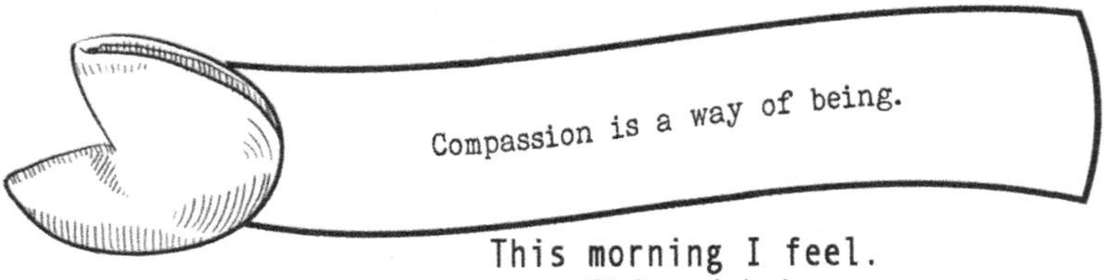

This morning I feel.
(Circle one below)

Fair

Displeased Exceptional

Displeased Exceptional

Fair

Tonight I feel.
(Circle one above)

One small victory for today was... ______________________________

____

What did I takeaway from today's challenge(s)? ______________________________

____

____

TIME M T W T F S S DATE:

TODAY, I AM GRATEFUL FOR... ______________________________

____

TO FEEL MORE ENERGIZED I WILL... ______________________________

AFFIRMATIONS, I AM... ______________________________

____

____

Now these three remaim, faith, hope, and love. The greatest of these is love.

This morning I feel.
(Circle one below)

Displeased Fair Exceptional

Displeased Fair Exceptional

Tonight I feel.
(Circle one above)

One small victory for today was... ______________________________

____

What did I takeaway from today's challenge(s)? ______________________________

____

____

TIME M T W T F S S DATE:

TODAY, I AM GRATEFUL FOR... ______________________________

____

____

TO FEEL MORE ENERGIZED I WILL... ______________________________

AFFIRMATIONS, I AM... ______________________________

____

____

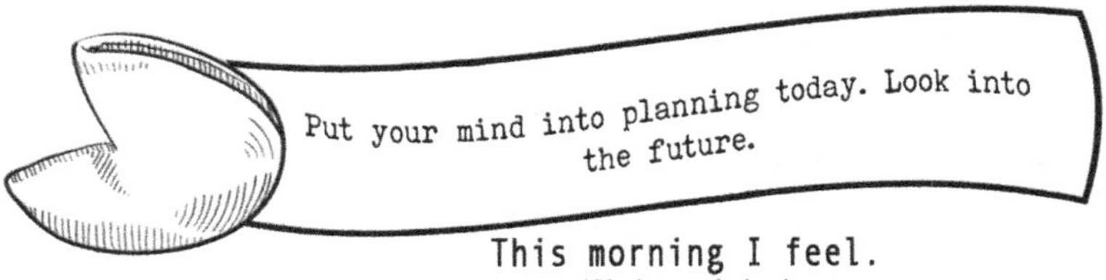

This morning I feel.
(Circle one below)

Fair

Displeased Exceptional

Displeased Exceptional

Fair

Tonight I feel.
(Circle one above)

One small victory for today was... ______________________________

____

What did I takeaway from today's challenge(s)? ______________________________

____

____

TIME M T W T F S S DATE:

TODAY, I AM GRATEFUL FOR... ____________________

____

TO FEEL MORE ENERGIZED I WILL... ____________________

AFFIRMATIONS, I AM... ____________________

____

____

The human spirit is stronger then anything that can happen to it..

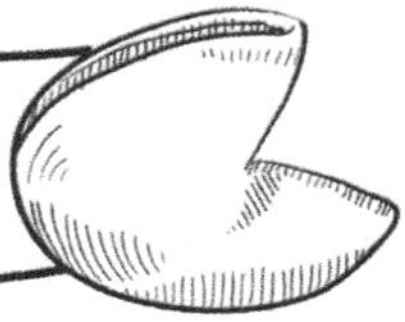

This morning I feel.
(Circle one below)

Fair

Displeased Exceptional

Displeased Exceptional

Fair

Tonight I feel.
(Circle one above)

One small victory for today was... ____________________

____

How can I make tomorrow better for myself and/or others? ____________________

____

____

TIME M T W T F S S DATE:

TODAY, I AM GRATEFUL FOR... ____________________

____

____

TO FEEL MORE ENERGIZED I WILL... ____________________

AFFIRMATIONS, I AM... ____________________

____

____

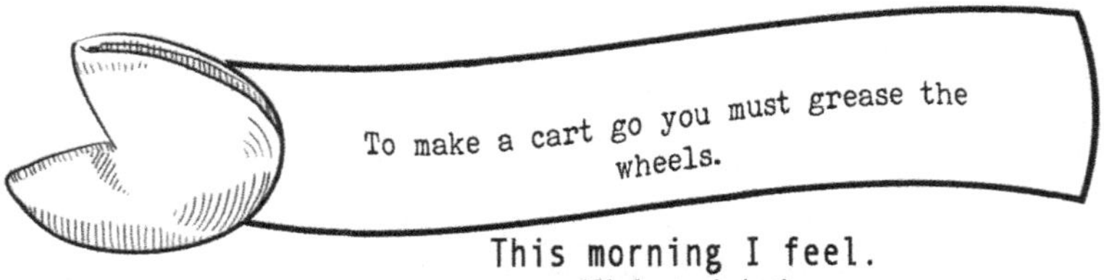

This morning I feel.
(Circle one below)

Fair

Displeased Exceptional

Displeased Exceptional

Fair

Tonight I feel.
(Circle one above)

One small victory for today was... ____________________

____

What is one small thing I can do for someone else next week?

____

Smile
MORE
worry
LESS

TIME M T W T F S S DATE:

TODAY, I AM GRATEFUL FOR...

TO FEEL MORE ENERGIZED I WILL...

AFFIRMATIONS, I AM...

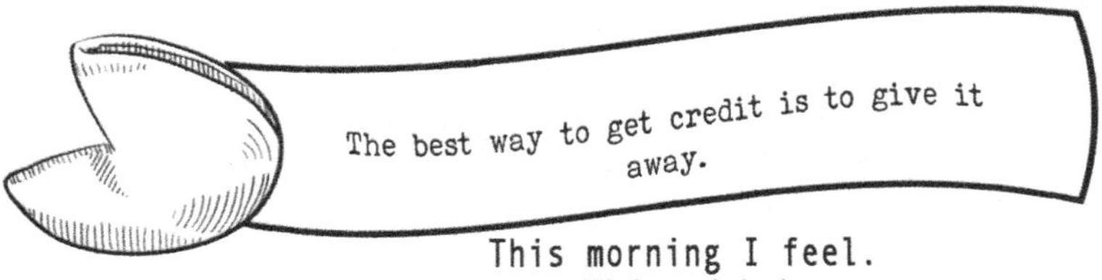

This morning I feel.
(Circle one below)

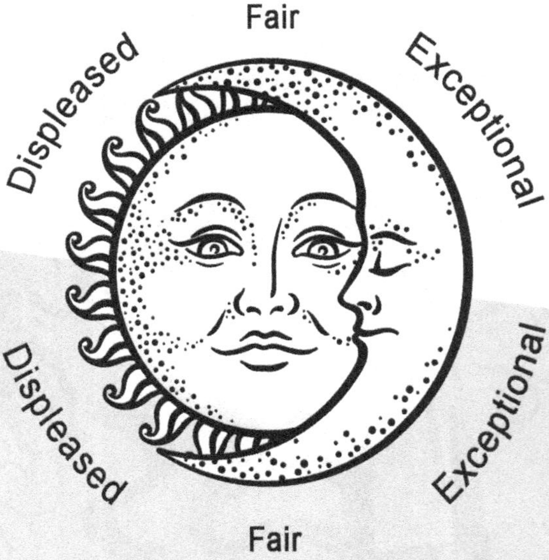

Tonight I feel.
(Circle one above)

One small victory for today was...

What did I takeaway from today's challenge(s)?

TIME M T W T F S S DATE:

TODAY, I AM GRATEFUL FOR... ____________________

TO FEEL MORE ENERGIZED I WILL... ____________________

AFFIRMATIONS, I AM... ____________________

A cheerful message is on its way to you.

This morning I feel.
(Circle one below)

Displeased Fair Exceptional

Displeased Fair Exceptional

Tonight I feel.
(Circle one above)

One small victory for today was... ____________________

What did I takeaway from today's challenge(s)? ____________________

TIME M T W T F S S DATE:

TODAY, I AM GRATEFUL FOR...

TO FEEL MORE ENERGIZED I WILL...

AFFIRMATIONS, I AM...

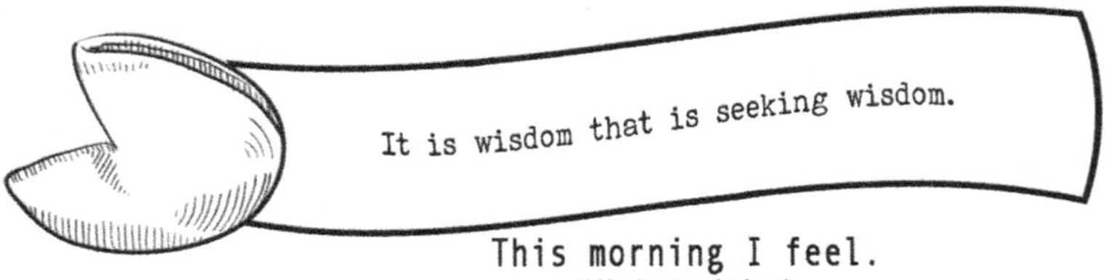

This morning I feel.
(Circle one below)

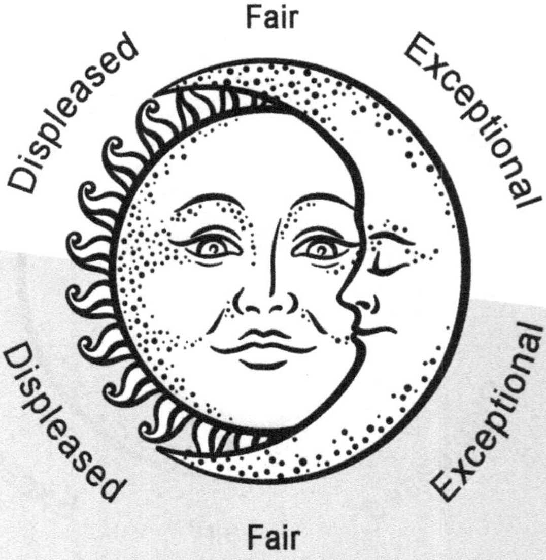

Tonight I feel.
(Circle one above)

One small victory for today was...

What did I takeaway from today's challenge(s)?

TIME M T W T F S S DATE:

TODAY, I AM GRATEFUL FOR... ______

______

TO FEEL MORE ENERGIZED I WILL... ______

AFFIRMATIONS, I AM... ______

______

______

Laughter is the shortest distance between two people.

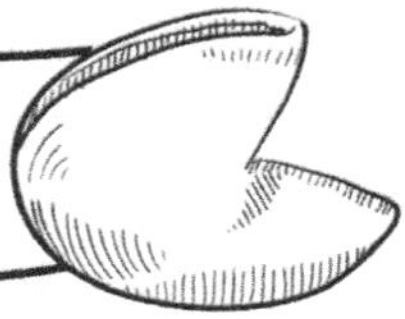

This morning I feel.
(Circle one below)

Fair

Displeased

Exceptional

Displeased

Exceptional

Fair

Tonight I feel.
(Circle one above)

One small victory for today was... ______

______

How can I make tomorrow better for myself and/or others? ______

______

______

TIME M T W T F S S DATE:

TODAY, I AM GRATEFUL FOR...

TO FEEL MORE ENERGIZED I WILL...

AFFIRMATIONS, I AM...

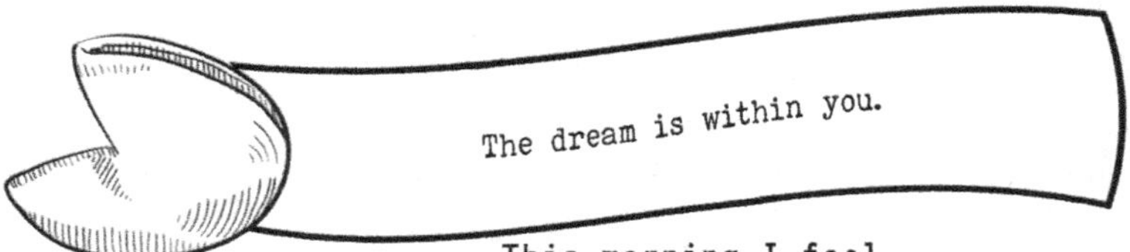

This morning I feel.
(Circle one below)

Fair

Displeased Exceptional

Displeased Exceptional

Fair

Tonight I feel.
(Circle one above)

One small victory for today was...

What is one small thing I can do for someone else next week?

the adventure begins

Write a small adventure you would like to take using the questions below.

Where?

When?

How to make it happen?

What do I need?

Who will go with?

TIME M T W T F S S DATE:

TODAY, I AM GRATEFUL FOR...

TO FEEL MORE ENERGIZED I WILL...

AFFIRMATIONS, I AM...

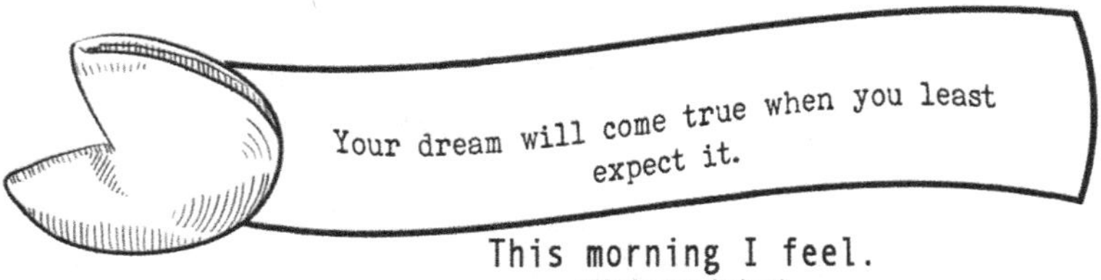

This morning I feel.
(Circle one below)

Fair

Displeased

Exceptional

Displeased

Exceptional

Fair

Tonight I feel.
(Circle one above)

One small victory for today was...

What did I takeaway from today's challenge(s)?

TIME M T W T F S S DATE:

TODAY, I AM GRATEFUL FOR... ______

______

TO FEEL MORE ENERGIZED I WILL... ______

AFFIRMATIONS, I AM... ______

______

______

Believing is doing.

This morning I feel.
(Circle one below)

Fair

Displeased

Exceptional

Displeased

Exceptional

Fair

Tonight I feel.
(Circle one above)

One small victory for today was... ______

______

What did I takeaway from today's challenge(s)? ______

______

______

TIME M T W T F S S DATE:

TODAY, I AM GRATEFUL FOR... ______________________

____

____

TO FEEL MORE ENERGIZED I WILL... ______________________

AFFIRMATIONS, I AM... ______________________

____

____

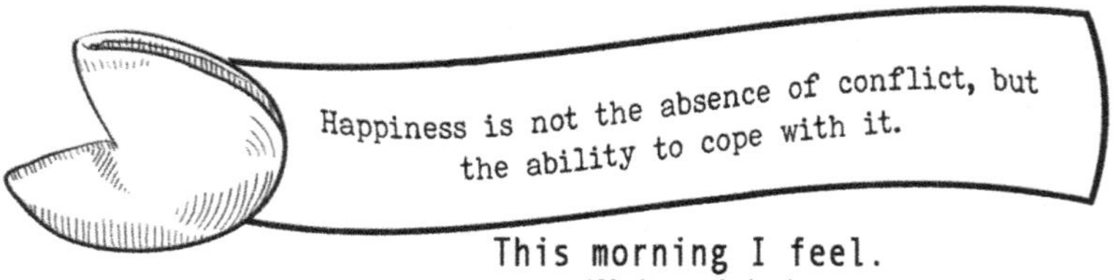

This morning I feel.
(Circle one below)

Fair

Displeased

Exceptional

Displeased

Exceptional

Fair

Tonight I feel.
(Circle one above)

One small victory for today was... ______________________

____

What did I takeaway from today's challenge(s)? ______________________

____

____

TIME M T W T F S S DATE:

TODAY, I AM GRATEFUL FOR...

TO FEEL MORE ENERGIZED I WILL...

AFFIRMATIONS, I AM...

You are far more influential than you think.

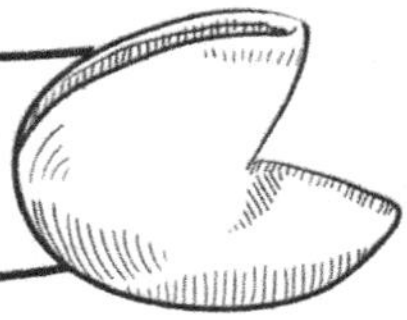

This morning I feel.
(Circle one below)

Displeased Fair Exceptional

Displeased Fair Exceptional

Tonight I feel.
(Circle one above)

One small victory for today was...

How can I make tomorrow better for myself and/or others?

TIME M T W T F S S DATE:

TODAY, I AM GRATEFUL FOR...

TO FEEL MORE ENERGIZED I WILL...

AFFIRMATIONS, I AM...

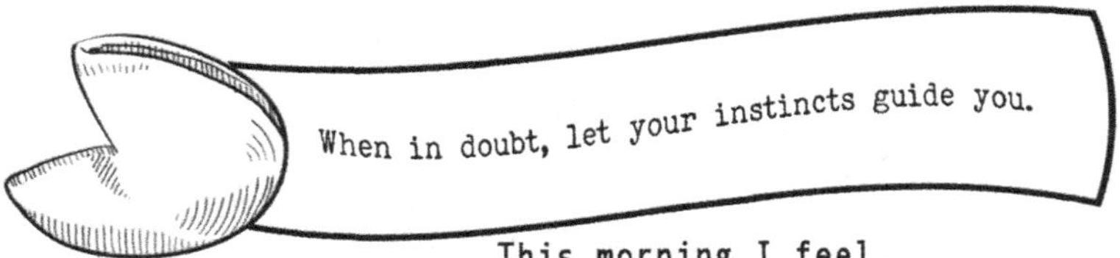

This morning I feel.
(Circle one below)

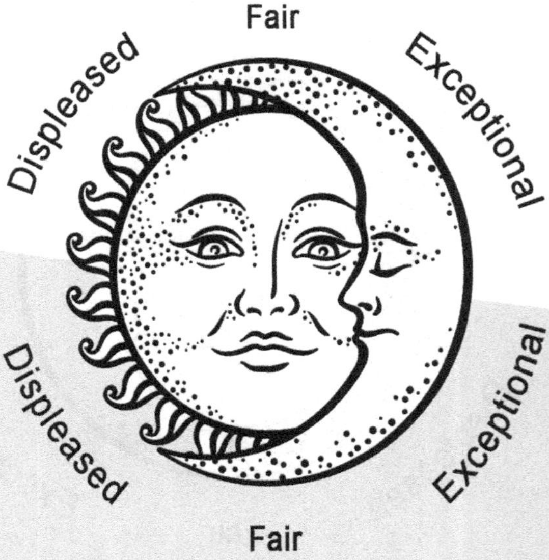

Tonight I feel.
(Circle one above)

One small victory for today was...

What is one small thing I can do for someone else next week?

the adventure begins

Write a small adventure you would like to take using the questions below.

Where?

When?

How to make it happen?

What do I need?

Who will go with?

TIME M T W T F S S DATE:

TODAY, I AM GRATEFUL FOR...

TO FEEL MORE ENERGIZED I WILL...

AFFIRMATIONS, I AM...

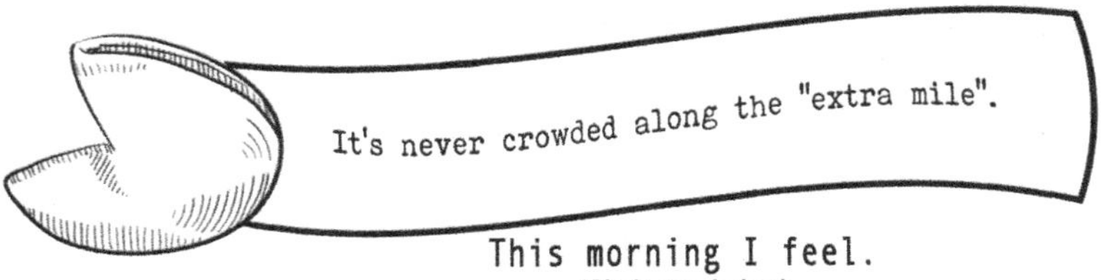

This morning I feel.
(Circle one below)

Displeased Fair Exceptional

Displeased Fair Exceptional

Tonight I feel.
(Circle one above)

One small victory for today was...

What did I takeaway from today's challenge(s)?

TIME M T W T F S S DATE:

TODAY, I AM GRATEFUL FOR...

TO FEEL MORE ENERGIZED I WILL...

AFFIRMATIONS, I AM...

Anyone who dares to be, can never be weak.

This morning I feel.
(Circle one below)

Displeased Fair Exceptional

Displeased Fair Exceptional

Tonight I feel.
(Circle one above)

One small victory for today was...

What did I takeaway from today's challenge(s)?

TIME M T W T F S S DATE:

TODAY, I AM GRATEFUL FOR... ______

______

______

TO FEEL MORE ENERGIZED I WILL... ______

AFFIRMATIONS, I AM... ______

______

______

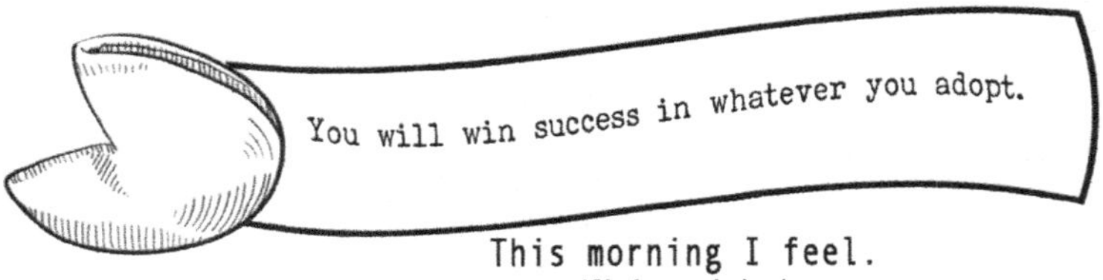

This morning I feel.
(Circle one below)

Fair

Displeased Exceptional

Displeased Exceptional

Fair

Tonight I feel.
(Circle one above)

One small victory for today was... ______

______

What did I takeaway from today's challenge(s)? ______

______

______

TIME M T W T F S S DATE:

TODAY, I AM GRATEFUL FOR... ______

______

TO FEEL MORE ENERGIZED I WILL... ______

AFFIRMATIONS, I AM... ______

______

______

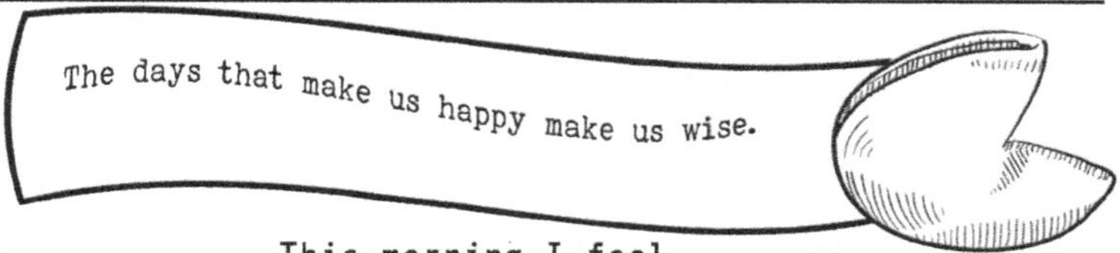

This morning I feel.
(Circle one below)

Displeased Fair Exceptional

Displeased Fair Exceptional

Tonight I feel.
(Circle one above)

One small victory for today was... ______

______

How can I make tomorrow better for myself and/or others? ______

______

______

TIME M T W T F S S DATE:

TODAY, I AM GRATEFUL FOR... ______________________________

____

____

TO FEEL MORE ENERGIZED I WILL... ______________________________

AFFIRMATIONS, I AM... ______________________________

____

____

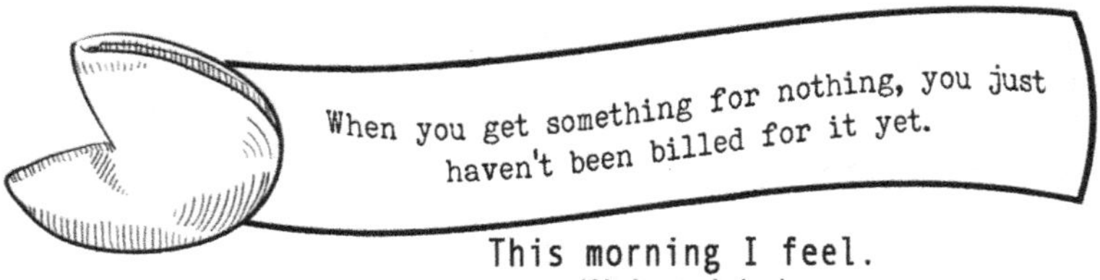

This morning I feel.
(Circle one below)

Tonight I feel.
(Circle one above)

One small victory for today was... ______________________________

____

What is one small thing I can do for someone else next week?

____

Smile
MORE
worry
LESS

TIME M T W T F S S DATE:

TODAY, I AM GRATEFUL FOR... ____________________

____

____

TO FEEL MORE ENERGIZED I WILL... ____________________

AFFIRMATIONS, I AM... ____________________

____

____

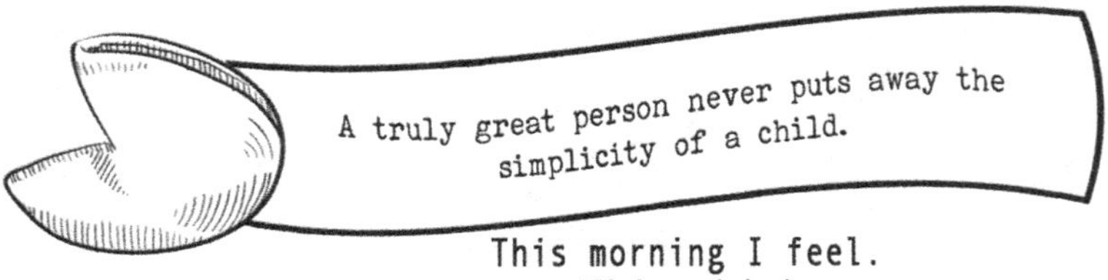

This morning I feel.
(Circle one below)

Fair

Displeased Exceptional

Displeased Exceptional

Fair

Tonight I feel.
(Circle one above)

One small victory for today was... ____________________

____

What did I takeaway from today's challenge(s)? ____________________

____

____

TIME M T W T F S S DATE:

TODAY, I AM GRATEFUL FOR... ______________________________

____

TO FEEL MORE ENERGIZED I WILL... ______________________________

AFFIRMATIONS, I AM... ______________________________

____

____

Believing that you are beautiful will make you appear beautiful to those around you.

This morning I feel.
(Circle one below)

Fair

Displeased

Exceptional

Displeased

Exceptional

Fair

Tonight I feel.
(Circle one above)

One small victory for today was... ______________________________

____

What did I takeaway from today's challenge(s)? ______________________________

____

____

TIME M T W T F S S DATE:

TODAY, I AM GRATEFUL FOR... ____________________

____

____

TO FEEL MORE ENERGIZED I WILL... ____________________

AFFIRMATIONS, I AM... ____________________

____

____

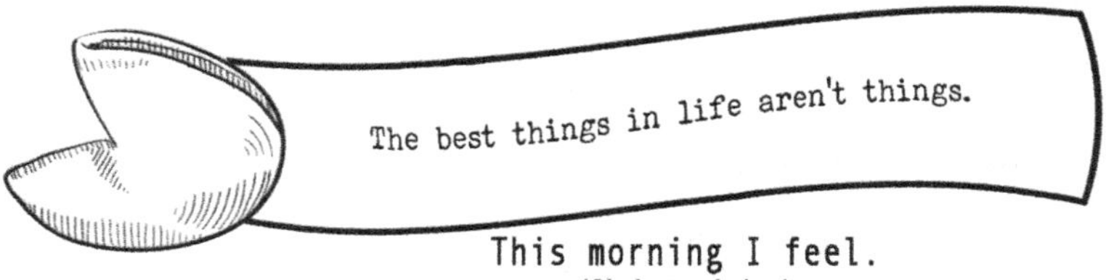

This morning I feel.
(Circle one below)

Fair

Displeased

Exceptional

Displeased

Exceptional

Fair

Tonight I feel.
(Circle one above)

One small victory for today was... ____________________

____

What did I takeaway from today's challenge(s)? ____________________

____

____

TIME M T W T F S S DATE:

TODAY, I AM GRATEFUL FOR...

TO FEEL MORE ENERGIZED I WILL...

AFFIRMATIONS, I AM...

It's OK to go against the grain.

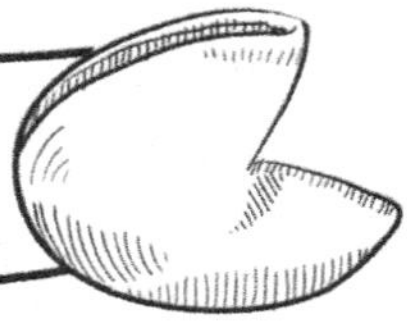

This morning I feel.
(Circle one below)

Displeased Fair Exceptional

Displeased Fair Exceptional

Tonight I feel.
(Circle one above)

One small victory for today was...

How can I make tomorrow better for myself and/or others?

TIME M T W T F S S DATE:

TODAY, I AM GRATEFUL FOR... ______

TO FEEL MORE ENERGIZED I WILL... ______

AFFIRMATIONS, I AM... ______

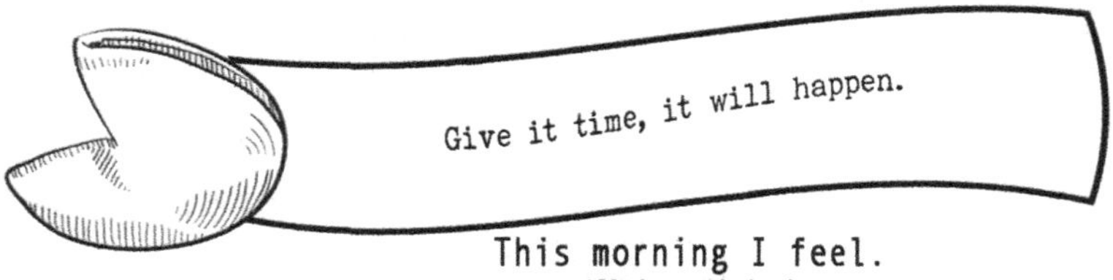

This morning I feel.
(Circle one below)

Fair

Displeased Exceptional

Displeased Exceptional

Fair

Tonight I feel.
(Circle one above)

One small victory for today was... ______

What is one small thing I can do for someone else next week?

Congratulations!

You made it to the end of the journal.

Are your desires the same as they were in the beginning?

Congratulations!

You have made it along way!

QR
Code
or

http://bit.ly/SeeUBeBrave

Paperback
JOURNAL